Copyright © by Harcourt, Inc.

All rights reserved. No part of this publication may be reproduced or transmitted in any form or by any means, electronic or mechanical, including photocopy, recording, or any information storage and retrieval system, without permission in writing from the publisher.

Requests for permission to make copies of any part of the work should be addressed to School Permissions and Copyrights, Harcourt, Inc., 6277 Sea Harbor Drive, Orlando, Florida 32887-6777. Fax: 407-345-2418.

HARCOURT and the Harcourt Logo are trademarks of Harcourt, Inc., registered in the United States of America and/or other jurisdictions.

Printed in México

ISBN-13: 978-0-15-362023-2

ISBN-10: 0-15-362023-4

6 7 8 9 10 0908 16 15 14 13 12

4500358761

Visit *The Learning Site!*
www.harcourtschool.com

Lesson 1

VOCABULARY

oxygen

What Do Animals Need to Live?

Oxygen is a gas found in air and water. All animals need oxygen in order to live. This bird gets oxygen from air.

READING FOCUS SKILL

MAIN IDEA AND DETAILS

The **main idea** is what the text is mostly about. **Details** tell more about the **main idea**.

Look for details about what animals need to live.

Animals and Their Needs

People care for pets. They make sure that pets have food, water, and a place to live. Wild animals have the same needs. But they must find their own food, water, and shelter.

Tell what all animals need to live.

◀ This panda finds food, water, and shelter in the wild.

▲ Hippos come to the water's surface to breathe air.

Animals Need Oxygen

All animals also need oxygen. **Oxygen** is a gas. It is in air and water. Many animals that live on land get oxygen by breathing air. Insects get oxygen through small holes in their bodies.

Many water animals, such as fish, get oxygen from water. Other water animals, such as whales, must rise to the top of water to get oxygen from air.

Name two ways that animals get oxygen.

Animals Need Food and Water

Animals need food to live and grow. Animals eat plants or other animlas to get food. They have body parts that help them get food.

Animals also need water. Most animals drink to get water. Some get water from food.

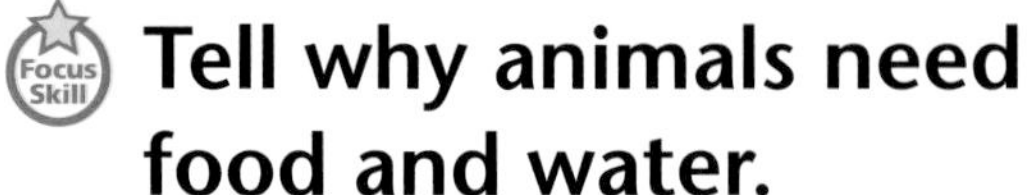

Tell why animals need food and water.

How do the body parts of these animals help them get food or water?

Animals Need Shelter

Animals need shelter. A shelter protects animals from weather. It also keeps them safe from other animals.

There are many kinds of shelters. Some birds build nests. Rabbits dig holes. This helps them hide. It keeps them safe.

▲ White tent bats find shelter under a leaf.

Why do animals need shelter?

Review

Focus Skill

Complete this main idea statement.

1. All animals need oxygen, food, ______ , and shelter.

Complete these detail sentences.

2. All animals get ______ from air or water.
3. All animals need ______ to get nutrients.
4. All animals need ______ to stay safe.

Lesson 2

What Are Vertebrates?

VOCABULARY

vertebrate
mammal
bird
reptile
amphibian
fish

A **vertebrate** is an animal with a backbone. This snake is a vertebrate.

A **mammal** is a vertebrate that has hair or fur. Most mammals give birth to live young.

A **bird** is a vertebrate that has feathers.

A **reptile** is a vertebrate that has dry skin covered with scales.

An **amphibian** is a vertebrate that has moist skin. Most have legs as adults.

A **fish** is a vertebrate that breathes through gills and lives in water.

READING FOCUS SKILL
MAIN IDEA AND DETAILS

The **main idea** is what the text is mostly about. **Details** tell more about the **main idea**.

Look for **details** about different kinds of vertebrates.

Vertebrates

A **vertebrate** is an animal that has a backbone. Scientists group vertebrates by the ways they are alike. There are five major groups of vertebrates. These groups are mammals, birds, reptiles, amphibians, and fish.

Tell what all vertebrates have.

How are these animals alike?

Mammals

A **mammal** is a vertebrate that has hair or fur. Mammals use lungs to breathe. Even mammals that live in water, such as whales, have lungs. They come to the surface of water to breathe air.

Most mammals give birth to live young. Mammal mothers make milk to feed their young.

▲ Pigs

Tell two ways all mammals are alike.

▼ Kangaroo

▼ Tiger

Birds

A **bird** is a vertebrate covered with feathers. Feathers can keep a bird warm. They also help some birds fly. Birds have a backbone and lungs.

Birds do not give birth to live young. They lay eggs. Young birds hatch from the eggs. Mother birds do not make milk to feed their young.

Tell three ways all birds are alike.

These birds look different, but they all have feathers, a backbone, and lungs.

Iguana ▶

Reptiles

A **reptile** is a vertebrate that has dry skin covered with scales. Most reptiles hatch from eggs laid on land. Like mammals and birds, reptiles use lungs to breathe.

Focus Skill **Tell three ways most reptiles are alike.**

▼ Alligator

▲ This frog began life as a tadpole.
Then it grew into a frog.

Amphibians

An **amphibian** is a vertebrate that has moist skin. Many amphibians have legs as adults. Most amphibians lay their eggs in water.

Frogs are amphibians. Their young are called tadpoles. Tadpoles hatch and live in water. They have gills instead of lungs. Gills take in oxygen from water. Tadpoles develop lungs as they grow. Then they can live on land. They also grow legs.

Tell two ways most amphibians are alike.

Fish

A **fish** is a vertebrate that breathes through gills. A fish lives in water.

Most fish lay eggs and have scales that cover their bodies. Scales are thin and strong. They help protect the fish. Fish use their fins to help them swim.

▼ **Swordfish**

Tell three ways most fish are alike.

Review

Focus Skill

Complete this main idea statement.

1. All ______ have a backbone.

Complete these detail statements.

2. ______ have hair or fur and feed milk to their young.
3. Both reptiles and ______ have bodies covered with scales.
4. ______ begin life with gills and then develop lungs to breathe air.

What Are Invertebrates?

VOCABULARY

invertebrate

An **invertebrate** is an animal without a backbone. A spider is an invertebrate.

READING FOCUS SKILL
COMPARE AND CONTRAST

When you **compare and contrast** you tell how things are alike and different.

Look for ways to **compare and contast** invertebrates.

Invertebrates

An **invertebrate** is an animal without a backbone. Most animals are invertebrates. There are more than a million kinds of invertebrates. Scientists group invertebrates by the ways they are alike.

Tell how invertebrates are different from vertebrates.

All these animals are invertebrates.

Insects

Insects are a kind of invertebrate. They do not have a backbone. There are more insects than any other kind of animal. Insects live everywhere on Earth.

Insects have three body parts and six legs. Insects have a hard outer covering. It helps keep them safe. Many insects have wings.

Tell how the animals on this page are alike.

Spiders and Ticks

Spiders and ticks may look like insects. But they are different. They have eight legs, not six. They have only two body parts.

Like insects, spiders and ticks have an outer body covering. This covering helps keep them safe.

Tell how spiders and ticks are the same as and different from insects.

▼ Deer tick

▼ Banana spider

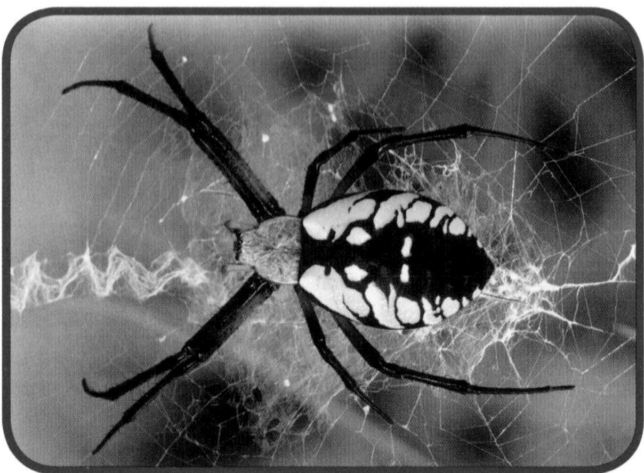

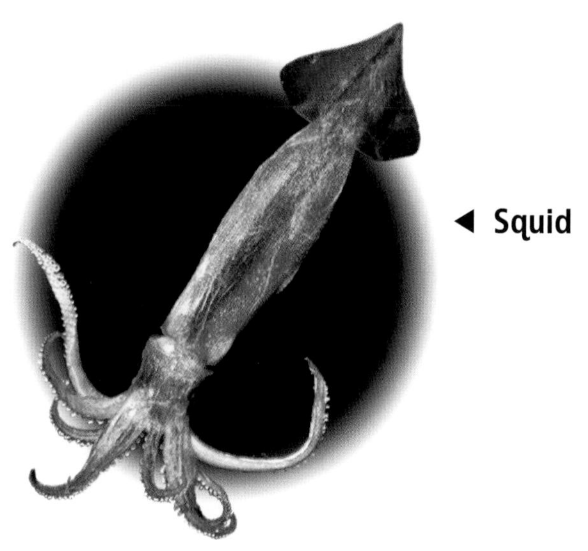

◀ Squid

Snails, Clams, and Squids

Snails, clams, and squids are in the same group of invertebrates. They all have soft bodies. Most animals in this group have a head. Some have hard shells to protect their bodies. Clams and snails have a kind of "foot." It helps them move.

How is the snail like the squid? How is it different?

Clams ▼

Snail ▶

Other Kinds of Invertebrates

Worms are another kind of invertebrate. They have no shells, legs, or eyes. Crabs and starfish are also invertebrates.

How is the crab like the starfish?

Starfish ▶

▼ Blue crab

▲ Earthworm

Invertebrates Are Important

Invertebrates are important to other living things. Bees move pollen from flower to flower. This helps plants make seeds. Earthworms mix air into soil.

Many invertebrates are food for other animals. People eat invertebrates, such as crabs. People also eat honey and use the wax made by bees.

▲ Honeybees

Tell how earthworms are like bees.

Review

Complete these sentences to compare and contrast invertebrates.

1. All invertebrates are animals without a ______.
2. There are more ______ than vertebrates on Earth.
3. Spiders and ticks have a different number of legs than ______.

GLOSSARY

amphibian (am•FIB•ee•uhn) A type of vertebrate that has moist skin—most have legs as an adult (14)

bird (BERD) A type of vertebrate that has feathers (12)

fish (FISH) A type of vertebrate that breathes through gills and spends its life in water (15)

invertebrate (in•VER•tuh•brit) An animal without a backbone (18)

mammal (MAM•uhl) A type of vertebrate that has hair or fur—most give birth to live young (11)

oxygen (AHK•si•juhn) A gas that animals need to live and that plants give off into the air (5)

reptile (REP•tyl) A type of vertebrate that has dry skin covered with scales (13)

vertebrate (VER•tuh•brit) An animal with a backbone (10)